MY NONFICTION DECODABLE READERS

Tap, Tip, Sap!

BY KIM THOMPSON

A Little Honey Book

Crabtree Publishing
crabtreebooks.com

Tips for Teachers and Caregivers

This book supports early readers as they decode words to learn facts and gain knowledge about the world.

Before reading, make sure students understand the sound-spelling correspondences shown below as well as the high-frequency words shown on the next page. Introduce the vocabulary words.

During reading, provide feedback and encouragement as students sound out decodable words by blending individual sounds.

After reading, talk about and write about the topic. Share the information on page 16 to help students learn more.

Letters and Sounds

New:

Sound	Spelling
short i	i
/p/	p

Review:

Sound	Spelling
short a	a
/m/	m
/s/	s
/t/	t

Decodable Words

at, is, it, map, sap, sip, tap, tip

High-Frequency Words

New: out, runs, this

Review: the

Vocabulary Words

cook

pail

tree

map

This is the map.

tree

This is the **tree**.

Tap. Tap. Tap.

This runs out.

It is sap!

sap

pail

This is the **pail**.

Tip it.

cook

Cook it.

Sip it.

Mmm!

Build Background Knowledge

People change the environment by using things they need and want. As temperatures rise in early spring, pressure builds inside sugar maple trees and causes their sap to flow. People drill holes in the trees and insert metal taps or tubes to collect the sap. Then, the sap is boiled down to make sweet maple syrup. Pancakes, anyone?

TAP, TIP, SAP!

Written by: Kim Thompson
Designed by: Rhea Magaro
Series Development: James Earley
Educational Consultant: Marie Lemke, M.Ed.

Photographs: All images from Shutterstock

Crabtree Publishing

crabtreebooks.com 800-387-7650

Printed in China/012024/FE20231222

Published in Canada
Crabtree Publishing
616 Welland Ave.
St. Catharines, Ontario
L2M 5V6

Published in the United States
Crabtree Publishing
347 Fifth Ave
Suite 1402-145
New York, NY 10016

Library and Archives Canada Cataloguing in Publication
Available at Library and Archives Canada

Library of Congress Cataloging-in-Publication Data
Available at the Library of Congress

Hardcover: 978-1-0398-4426-1
Paperback: 978-1-0398-4508-4
Ebook (pdf): 978-1-0398-4585-5
Epub: 978-1-0398-4655-5
Read-Along: 978-1-0398-4725-5
Audio: 978-1-0398-4795-8